I0106094

AI ADVENTURES

A 5th Grader's Guide to Mastering ChatGPT & Beyondy

"Unlock the Power of Artificial Intelligence to Transform Your Learning Experiencely

"Explore. Learn. Create. Discover the future of AI, one step at a time."

TURTLE

AI Adventures: A 5th Grader's Guide to Mastering ChatGPT & Beyond

Copyright © 2023 Turtle
All rights reserved. No part of this book may be reproduced or transmitted in any form or by any means without the written permission of the author.

Published by:
Eleviv Publishing Group
Centerville, OH 45458
info@elevivpublishing.com
www.elevivpublishing.com

ISBN: 978-1-952744-72-3 (PB)
       978-1-952744-73-0 (HC)
       978-1-952744-74-7 (E-book)

Printed in the United States of America

Welcome to the world of artificial intelligence, where the future is now! "AI Adventures: A 5th Grader's Guide to Mastering ChatGPT & Beyond" is the perfect companion for curious minds eager to harness the power of AI and transform their learning experience.

In this easy-to-follow guide, young learners will explore the fascinating world of AI, discover how to use ChatGPT effectively, and unlock the potential of AI-powered tools tailored just for them. With engaging examples, exciting activities, and a wealth of kid-friendly resources, this book aims to inspire the next generation of AI enthusiasts.

Inside, you'll find:
- An introduction to AI and ChatGPT
- Tips and tricks for crafting effective prompts
- A collection of creative prompts to spark your imagination
- A list of AI tools and websites designed for young learners
- A glimpse into the future of AI and the exciting careers it offers
- 100 fun prompts you can use
- Fun Games and Interactive activities and much more

Embark on an unforgettable journey through the captivating world of AI, where learning is made fun, interactive, and personalized. Join us in "AI Adventures" and take your first steps toward becoming an AI-savvy 5th grader today!

Turtle is a platform creating curiosity and wonder by creating tech learning tools, curriculums, study guides, books, and websites for children ages 1 to 18.

TABLE OF CONTENTS:

1. Introduction 10

2. What is AI? 13

3. Meet ChatGPT 17

4. The Power of Prompts 20

5. Crafting Your Own Prompts 24

6. Understanding AI Limitations 28

7. Safety Tips for Chatting with AI 32

8. Fun AI Activities 36

9. Conclusion 40

OVERVIEW

Chapter 1: Introduction

Welcome, young explorer! In this e-book, we're going to learn all about Artificial Intelligence (AI), ChatGPT, and the world of prompts. By the end of this adventure, you'll know how to create your own prompts and understand the exciting world of AI. So, buckle up and let's get started!

Chapter 2: What is AI?

AI, or Artificial Intelligence, is a branch of computer science that focuses on creating machines that can think, learn, and solve problems like humans do. These machines, called computers or robots, use algorithms and data to improve their skills. Some examples of AI are self-driving cars, voice assistants like Siri or Alexa, and even video games with characters that learn from their actions.

Chapter 3: Meet ChatGPT

ChatGPT is an AI language model created by OpenAI. It's designed to understand and generate human-like text. It's kind of like having a virtual friend that can chat with you, answer questions, and help you with your homework. ChatGPT is trained on lots of text data from books, articles, and websites to understand human language and respond to prompts.

Chapter 4: The Power of Prompts

A prompt is a message, question, or statement that you give to ChatGPT to get a response. Prompts help guide the AI in understanding what you want to talk about. For example, if you ask ChatGPT, "What is the capital of France?" it will understand that you want to know about the capital city and respond with "Paris."

Chapter 5: Crafting Your Own Prompts

Creating good prompts is an important skill when talking to ChatGPT. Here are some tips to help you craft effective prompts:

1. Be clear and specific: Make sure your prompt is easy to understand.
2. Ask open-ended questions: These encourage more detailed responses.
3. Experiment: Try different prompts to see what works best.

Let's practice! Imagine you want to know about the life cycle of a butterfly. Instead of asking, "Tell me about butterflies," you could ask, "Can you explain the life cycle of a butterfly?" This prompt is more specific and will give you a better answer.

Chapter 6: Understanding AI Limitations
AI like ChatGPT is impressive, but it's not perfect. It's important to know its limitations:
1. It might not always understand your prompt.
2. Sometimes it gives incorrect or outdated information.
3. It can be sensitive to how you phrase your prompts.
4. It can generate long responses that may not answer your question directly.

Keep these limitations in mind while chatting with ChatGPT and always double-check important information.

Chapter 7: Safety Tips for Chatting with AI
When talking to AI like ChatGPT, it's essential to stay safe and protect your privacy. Here are some tips:

1. Never share personal information, like your full name, address, or phone number.
2. Don't ask the AI for harmful or dangerous information.
3. Remember that AI doesn't have feelings, so treat it like a tool, not a friend.

Chapter 8: Fun AI Activities
Now that you know about prompts and AI, let's try some fun activities with ChatGPT!

1. Homework Helper: Ask ChatGPT to help you with your homework questions or explain difficult concepts.

2. Storytime: Create a story together! Give ChatGPT a story prompt, like "Once upon a time in a magical forest," and let it generate the next sentences. Keep adding to the story until you have an exciting tale.
3. Virtual Travel Guide: Ask ChatGPT about different countries, their cultures, and famous landmarks. You can even ask for recommendations on what to do in a specific city or region.
4. Riddles and Puzzles: Test ChatGPT's problem-solving skills by giving it riddles or puzzles to solve. You can also ask it to create new riddles for you to solve.
5. Art Ideas: If you love drawing or painting, ask ChatGPT for ideas on what to create. You can give it themes or let it suggest something completely new.

Chapter 9: Conclusion

Congratulations, you've reached the end of our AI adventure! Now you know what AI and ChatGPT are, how to use prompts, and some fun activities you can try. Remember to stay safe while chatting with AI and always be curious about learning new things. With your new AI knowledge, you're ready to explore the world of ChatGPT and create amazing conversations. Happy chatting!

Chapter 1

INTRODUCTION

Welcome to the fantastic world of Artificial Intelligence, young explorer! This book is designed to help you, a bright 5th grader, learn all about AI, ChatGPT, and the power of prompts. As you journey through these chapters, you'll uncover the secrets of crafting great prompts and learn how to use AI responsibly and safely.

We'll dive into the fascinating world of ChatGPT, an AI language model created by OpenAI, and explore how it understands and generates human-like text. By the end of this adventure, you'll be a whiz at creating your own prompts and navigating the exciting world of AI. So, put on your thinking cap, and let's get started on this amazing journey!

You might be wondering what AI is and how it affects our lives. AI is a part of computer science that focuses on building machines and software that can think, learn, and make decisions like humans do. These smart machines use algorithms, which are sets of step-by-step instructions that help them complete tasks or solve problems.

Over the years, AI has become an essential part of our daily lives. It's in our phones, helping us find the quickest route to school; it's in our video games, creating challenging and fun experiences; and it's even in our homes, with smart devices that can play our favorite songs or turn off the lights when we ask.

As AI continues to advance, it's important for young learners like you to understand how it works and how you can make the most of it. That's where ChatGPT comes in!

So what is ChatGPT?
ChatGPT is a powerful AI language model developed by OpenAI. It's designed to understand and generate human-like text based on the prompts it receives. A prompt is a question, statement, or instruction that guides the AI's response or action. By giving ChatGPT clear and effective prompts, you can have engaging and informative conversations with the AI.

ChatGPT is a great tool for learning, creativity, and exploration. You can use it to get help with your homework, brainstorm ideas for a story, or even learn fun facts about a topic you're interested in. The possibilities are endless!

In this book, we'll help you understand how to use ChatGPT

effectively, explore various AI tools and resources designed for kids, and discover the exciting future of AI. We've also included fun activities and games that you can enjoy as you learn more about AI and ChatGPT. So, are you ready to begin your AI adventure? Let's dive in!

Chapter 2

WHAT IS AI?

AI, or Artificial Intelligence, is a field in computer science that focuses on creating machines or computer programs that can think, learn, and solve problems like humans. AI systems can analyze large amounts of data, recognize patterns, and make decisions based on that information. They continue to improve their skills and knowledge by learning from their experiences.

There are many different types of AI that you might encounter in your daily life. Some examples include self-driving cars, voice assistants like Siri or Alexa, and even video games with characters that learn from your actions. These AI systems use complex algorithms and lots of data to become smarter over time, helping them to better understand and interact with the world around them.

a. The Concept of Artificial Intelligence

Artificial Intelligence, or AI, is a branch of computer science that aims to create machines or software programs that can perform tasks that would typically require human intelligence. This involves the development of algorithms and models that enable computers to learn, reason, and make decisions.

Machine Learning: Machine Learning is a subset of AI that focuses on creating systems that can learn from data and improve their performance over time. These systems use algorithms that can recognize patterns, make predictions, and adapt their behavior based on new information.

Neural Networks: Inspired by the structure and function of the human brain, neural networks are a type of machine learning model that can process complex information and make decisions based on that information. They are the foundation for many advanced AI systems, like ChatGPT.

b. Types of AI

AI can be broadly categorized into two types: narrow AI and general AI.

Narrow AI: Also known as weak AI, narrow AI refers to AI systems that are designed to perform specific tasks or solve particular problems. Examples include recommendation systems, virtual assistants like Siri and

Alexa, and AI tools like ChatGPT. Narrow AI systems are highly specialized and excel at their designated tasks but cannot perform tasks outside their domain.

General AI: Also known as strong AI, general AI refers to hypothetical AI systems that possess the ability to understand, learn, and perform any intellectual task that a human being can do. General AI does not exist yet, and its development remains a subject of ongoing research and debate.

c. AI in Everyday Life

AI has become an integral part of our daily lives, even if we don't always realize it. From search engines that help us find information to social media algorithms that curate our feeds, AI systems are constantly working behind the scenes to enhance our experiences.

Personal Assistants: Virtual assistants like Siri, Alexa, and Google Assistant use AI to understand and respond to our voice commands, helping us manage our schedules, answer questions, or control smart home devices.

Entertainment: Recommendation algorithms on platforms like Netflix, YouTube, and Spotify use AI to analyze our preferences and suggest content we might enjoy.

Education: AI-powered tools like ChatGPT can assist with learning by providing explanations, solving problems, or offering creative inspiration.

By understanding the concept of AI, its types, and its role in our everyday lives, you can better appreciate the potential and value of AI tools like ChatGPT and use them to enhance your learning, creativity, and enjoyment.

Chapter 3

MEET CHATGPT

ChatGPT is an AI language model developed by OpenAI. Its main purpose is to understand and generate human-like text, making it an incredibly powerful tool for communication, learning, and creativity. Think of ChatGPT as a virtual friend that can chat with you, answer questions, and even help you with your homework!

To become this smart, ChatGPT has been trained on a vast amount of text data from books, articles, and websites. It uses this knowledge to understand the context and meaning behind the words you use in your prompts. By processing and analyzing the text data, ChatGPT can generate intelligent and relevant responses to your questions and statements.

Examples of what you can do with ChatGpt. ChatGPT is a

versatile AI tool that can assist you in numerous tasks and activities. Here are some examples of what you can do with ChatGPT:

Homework Help: ChatGPT can help you understand complex concepts, solve math problems, or provide explanations for science, history, and other subjects.

Creative Writing: Use ChatGPT to generate story ideas, character descriptions, or even write entire passages to help you with your creative writing projects.

Brainstorming: Ask ChatGPT to help you come up with ideas for school projects, presentations, or essays by generating suggestions or outlining key points.

Language Learning: ChatGPT can assist you in learning new languages by providing translations, helping you practice conversation, or explaining grammar rules.

Trivia and Facts: Use ChatGPT to learn interesting facts or answer trivia questions on a wide range of topics.

Art and Design Inspiration: ChatGPT can provide creative prompts or descriptions for drawing, painting, or other artistic projects.

Debates and Perspectives: Ask ChatGPT to explain different viewpoints on a topic, helping you understand various perspectives and form your own opinions.

Jokes and Riddles: Have fun with ChatGPT by asking it to tell jokes, riddles, or create humorous stories.

Recommendations: ChatGPT can provide recommendations for books, movies, games, or other forms of entertainment based on your interests.

Relaxation and Meditation: ChatGPT can generate guided meditation scripts, relaxation techniques, or mindfulness practices to help you unwind and find peace.

Remember that while ChatGPT can be a valuable resource, it's essential to verify the information it provides and use it as a starting point for further research or exploration. By experimenting with different prompts and interactions, you can discover countless ways to engage with ChatGPT and enhance your learning, creativity, and enjoyment.

Chapter 4

THE POWER OF PROMPTS

Prompts are the key to unlocking ChatGPT's potential. A prompt is a message, question, or statement that you give to ChatGPT to guide its response. By providing clear and specific prompts, you can help the AI understand your needs and generate more accurate and helpful answers.

For example, if you ask ChatGPT a vague question like, "What's that thing in the sky?" it might not know exactly what you're referring to. However, if you ask, "What are the different types of clouds in the sky?" the AI will have a much better understanding of what you want to know, and it can provide a more detailed response.

Prompts are the foundation of your interactions with ChatGPT, so learning how to create effective prompts is essential for getting the most out of your AI conversations.

a. The Role of Prompts in AI Conversations

Prompts play a crucial role in guiding AI interactions and determining the quality of the responses you receive. By providing clear, concise, and engaging prompts, you can unlock the full potential of AI tools like ChatGPT and create meaningful conversations.

Starting a Conversation: A well-crafted prompt can initiate an interesting conversation with ChatGPT, helping you explore new topics, ideas, or perspectives.

Guiding the Discussion: Your prompts can help shape the direction of the conversation, allowing you to focus on specific subjects or delve deeper into particular aspects of a topic.

Inspiring Creativity: Creative prompts can challenge ChatGPT to generate unique ideas, stories, or suggestions, sparking your imagination and encouraging you to think outside the box.

b. The Benefits of Prompt Engineering

Prompt engineering is the skill of crafting effective prompts that yield desired responses from AI tools. Mastering prompt engineering can provide numerous benefits, including:

Improved Communication: Learning to write clear and concise prompts can improve your overall communication skills, helping you express your thoughts and ideas more effectively.

Enhanced Problem Solving: By crafting prompts that encourage ChatGPT to generate solutions or suggestions, you can enhance your problem-solving abilities and explore multiple approaches to challenges.

Boosted Creativity: Developing the skill of prompt engineering can inspire you to think creatively and come up with original ideas, both in your AI interactions and in other aspects of your life.

c. The Value of Experimentation

Experimentation is key to mastering the power of prompts. By trying out different prompts and observing the responses from ChatGPT, you can learn what works best and refine your prompt engineering skills.

Test Different Approaches: Experiment with various types of prompts, such as questions, statements, and instructions, to discover which approach yields the most satisfying results.

Refine Your Prompts: If a prompt doesn't generate the desired response, try rephrasing it or adding more context to improve the outcome.

Learn from Successes and Failures: Pay attention to the prompts that work well and those that don't. Analyze your successes and failures to better understand the factors that contribute to effective prompting.

By harnessing the power of prompts and embracing the value of experimentation, you can unlock the full potential of AI tools like ChatGPT and make your AI interactions more engaging, informative, and enjoyable.

Chapter 5

CRAFTING YOUR OWN PROMPTS

Now that you understand the importance of prompts, let's learn how to craft your own! Creating effective prompts is both an art and a science, but with a bit of practice, you'll be a master in no time. Here are some helpful tips for crafting great prompts:

1. **_Be clear and specific:_** Make sure your prompt is easy to understand and leaves little room for confusion. The more specific you are, the better ChatGPT can tailor its response to your needs.

2.   *Ask open-ended questions:* Open-ended questions encourage more detailed and thoughtful responses from the AI. Instead of asking yes or no questions, try asking questions that start with "how," "why," or "what."

3.   *Experiment:* Don't be afraid to try different prompts to see what works best. You might be surprised by the variety of responses you can get from ChatGPT with just a slight change in your prompt.

For example, let's say you want to learn about the planets in our solar system. Instead of asking a simple question like, "What are the planets?" you could ask, "Can you provide a brief description of each planet in our solar system?" This more specific prompt will encourage ChatGPT to give you a more detailed and informative response.

Crafting Effective Prompts
a. Understanding Prompts

Prompts are the key to unlocking the full potential of ChatGPT. A prompt is a question, statement, or instruction that guides the AI's response or action. When you give ChatGPT a prompt, the AI processes it and generates a relevant response based on the information it has learned from the billions of texts it has been trained on. The better your prompt, the more helpful and accurate the AI's response will be. To create effective prompts, it's important to understand the types of prompts you can use:

Questions: You can ask ChatGPT questions on various topics, like science, history, or math. For example, "What are the three states of matter?" or "Who was the first person to walk on the moon?"

Statements: Provide a statement or a topic to explore, and ChatGPT will generate a response based on it. For example, "Tell me about photosynthesis" or "Describe life in ancient Egypt."

Instructions: You can give ChatGPT specific tasks or requests. For example, "Write a short story about a talking cat" or "Help me come up with ideas for a science project."

b. Tips for Writing Good Prompts

When crafting prompts for ChatGPT, keep these tips in mind: Be clear and specific: Make sure your prompt is easy to understand and provides enough information for ChatGPT to generate a relevant response.

Use keywords: Include important keywords related to your topic or question to help ChatGPT understand the context and provide accurate information. Set a format: If you want ChatGPT to present information in a particular way, specify the format in your prompt. For example, "List the five largest planets in our solar system in order from largest to smallest."

Limit the response length: If you want a shorter or

longer response, you can specify the desired length in your prompt. For example, "Explain the water cycle in three sentences." Experiment and iterate: Sometimes, ChatGPT may not provide the exact response you were looking for. Don't be afraid to rephrase your prompt or ask a follow-up question to get the information you need.

c. Common Mistakes to Avoid

When using ChatGPT, it's essential to avoid certain mistakes that could lead to unhelpful or inaccurate responses:

Vague or ambiguous prompts: If your prompt is too vague or unclear, ChatGPT may not understand what you're asking or provide irrelevant information. Overloading the prompt: Avoid including too many details or requests in a single prompt, as this might confuse the AI or result in an incomplete response. Relying solely on AI: While ChatGPT is an incredible tool, it's essential to verify the information it provides and use it as a starting point for further research or exploration.

By following these tips and avoiding common mistakes, you'll be able to craft effective prompts that help you get the most out of your interactions with ChatGPT.

Chapter 6

UNDERSTANDING AI LIMITATIONS

As amazing as AI and ChatGPT can be, it's crucial to remember that they are not perfect. Understanding the limitations of AI will help you have more meaningful and productive interactions with ChatGPT. Here are some important limitations to keep in mind:

1. ***Misunderstanding prompts:*** ChatGPT might not always understand your prompt correctly, which can lead to confusing or irrelevant responses.

2. ***Incorrect or outdated information:*** ChatGPT might sometimes provide incorrect or outdated information, as its knowledge is based on the text data it has been

trained on.

3.   *Sensitivity to phrasing:* The way you phrase your prompt can have a significant impact on the AI's response, so it's essential to be mindful of your wording.

4.   *Lengthy responses:* ChatGPT might generate long responses that may not directly answer your question or address your prompt.

When using ChatGPT, always be aware of these limitations and double-check important information to ensure accuracy.

AI Is Not Perfect

As amazing as AI and ChatGPT can be, it's essential to understand that they have limitations. AI is not perfect, and sometimes it may provide inaccurate information or misunderstand your prompt. Recognizing these limitations will help you use AI tools more effectively and responsibly.

Limited Knowledge: AI models like ChatGPT are trained on vast amounts of data from various sources, but their knowledge is limited to what they've been trained on. If new information or discoveries have emerged since their last update, the AI may not be aware of them.

Bias: AI models can sometimes reflect the biases present in the data they've been trained on. This can lead to unfair or stereotypical responses that may not accurately represent diverse perspectives or experiences.

Lack of Common Sense: While AI models can understand language and context to some extent, they may lack common sense or the ability to reason like humans. This can result in responses that don't make sense or aren't practical.

b. Strategies for Overcoming Limitations

To make the most of your AI experience, consider these strategies for overcoming some of the limitations:

Verify Information: When ChatGPT provides you with information, it's a good idea to verify it using trusted sources like textbooks, educational websites, or by asking a knowledgeable adult. This will help ensure you're getting accurate and up-to-date information.

Reframe Prompts: If you receive an unsatisfactory or confusing response from ChatGPT, try rephrasing your prompt or asking a follow-up question to get a better answer.

Discuss AI Ethics: Engage in conversations about the ethical implications of AI and its potential biases. Understanding these issues can help you make more informed decisions when using AI tools like ChatGPT.

c. Embracing AI Responsibly

As a 5th grader learning about AI and ChatGPT, it's important to embrace these powerful tools responsibly. Remember that AI is not a substitute for human judgment, creativity, or empathy. Instead, it's a valuable resource that can complement and enhance your learning and exploration. By understanding AI's limitations, you'll be better equipped to use ChatGPT effectively and responsibly, making your AI adventure even more exciting and rewarding.

Chapter 7

SAFETY TIPS FOR CHATTING WITH AI

When interacting with AI like ChatGPT, it's essential to prioritize safety and protect your privacy. Here are some important safety tips to follow while using AI:

1. Never share personal information: Avoid sharing personal details like your full name, address, or phone number. This helps protect your privacy and ensures your conversations with AI remain safe.

2. Don't ask for harmful or dangerous information: Be responsible when using AI and avoid asking for information that could be harmful or dangerous.

3. *Treat AI as a tool, not a friend:* Remember that AI, like ChatGPT, doesn't have feelings or emotions. It's a powerful tool for learning and exploration, but it's not a substitute for real friendships and human connections.

By following these safety tips, you can ensure that your interactions with AI remain fun, educational, and secure.

a. Protecting Personal Information

When interacting with AI tools like ChatGPT, it's essential to keep your personal information safe. Sharing personal details online can expose you to potential risks. To protect yourself, always follow these safety tips:

Never share personal information: Avoid sharing your full name, address, phone number, school, or any other identifying details when using AI tools.

Use a nickname or initials: If you need to provide a name when interacting with ChatGPT, use a nickname or your initials instead of your real name.

Be cautious with photos and videos: Don't share personal photos or videos with AI tools, as they might be stored or used in ways you don't expect.

b. Monitoring AI Interactions

To ensure a positive and safe experience while using AI tools, follow these guidelines:

Use age-appropriate AI tools: Make sure you're using AI tools and platforms that are designed for kids your age. These tools often have built-in safety features and content filters to create a kid-friendly environment.

Parental supervision: It's a good idea to involve your parents or guardians in your AI interactions. They can help you understand how to use AI tools responsibly and ensure your safety while engaging with them.

Report inappropriate content: If you encounter any inappropriate content or experience something that makes you uncomfortable while using an AI tool, notify your parents or guardians and report the issue to the platform's support team.

c. Encouraging Positive AI Interactions

As you explore the world of AI, it's important to foster positive interactions and use AI tools responsibly:

Be respectful: Treat AI tools with respect, just as you would with a human. Using polite language and avoiding offensive or inappropriate content helps create a positive environment for everyone.

Learn and grow: Use AI tools as an opportunity to learn, explore, and expand your knowledge. Ask thoughtful questions and engage in meaningful conversations to make the most of your AI experience.

Share your experiences: Talk to your friends, family, and classmates about your AI interactions. Sharing your experiences can help others learn about AI and foster a supportive community of responsible AI users.

By following these safety tips and promoting positive AI interactions, you can ensure a safe and enjoyable experience while chatting with AI tools like ChatGPT.

Chapter 8

FUN AI ACTIVITIES

Now that you're equipped with the knowledge of prompts and AI, it's time to have some fun with ChatGPT! Here are a few exciting activities to try:

1. Homework Helper: Ask ChatGPT to help you with your homework questions or explain challenging concepts in a way that's easy to understand.

2. Storytime: Collaborate with ChatGPT to create a story. Give the AI a story prompt, and let it generate the next few sentences. Keep building on the story until you have an entertaining tale to share with friends and family.

3. Virtual Travel Guide: Learn about different countries, cultures, and famous landmarks by asking ChatGPT for information and recommendations on various destinations.

4. Riddles and Puzzles: Test ChatGPT's problem-solving skills by giving it riddles or puzzles to solve. You can also ask it to create new riddles for you to figure out.

5. Art Inspiration: If you enjoy drawing or painting, ask ChatGPT for ideas on what to create. You can provide themes or let the AI suggest something completely new and unique.

By exploring these activities, you'll learn more about the capabilities of AI and develop your skills in crafting effective prompts.

a. Creative Writing with AI

AI tools like ChatGPT can be a great source of inspiration for your creative writing. Here are some fun activities to help you boost your creativity:

Story Starters: Provide ChatGPT with a sentence or a scene to start a story and let the AI generate the next few paragraphs. You can then continue the story or ask the AI to provide more ideas.

Character Creation: Ask ChatGPT to help you create

unique characters for your stories. Provide a few details like name, age, and personality traits, and let the AI generate a character description or backstory.

Dialogue Practice: Write a conversation between two characters, and let ChatGPT generate responses for one or both characters. This can help you practice writing dialogue and create engaging interactions between characters.

b. AI-Powered Homework Help

ChatGPT can be a valuable resource to help you with your homework:

Math Assistance: If you're stuck on a math problem, you can ask ChatGPT for help. Provide the problem and let the AI guide you through the steps to solve it.

Science Explanations: If you're having trouble understanding a science concept, ask ChatGPT to explain it in simple terms or provide real-life examples to help you grasp the idea better.

Essay Outlines: When writing an essay, you can ask ChatGPT to help you create an outline or suggest some key points to include in your paper.

c. AI-Generated Art Projects

You can use AI tools to create fun and unique art projects:

AI-Powered Drawing Prompts: Ask ChatGPT to generate a unique drawing prompt or describe a scene for you to illustrate. This can help you practice your drawing skills and challenge your creativity.

AI-Generated Stories for Comics: Create a comic strip or graphic novel using AI-generated stories. Provide ChatGPT with a theme or a setting, and let the AI generate a storyline for you to illustrate.

Collaborative AI Art: Work with your friends to create a collaborative art piece using AI. Each person can ask ChatGPT for a unique element or character to include in the artwork, and then combine your drawings to create a collaborative masterpiece.

By engaging in these fun AI activities, you can harness the power of AI to spark your creativity, enhance your learning, and explore new ideas.

Chapter 9

CONCLUSION

Congratulations, young explorer! You've reached the end of our AI adventure, and now you're well-equipped with the knowledge and skills to interact with AI like ChatGPT. You've learned about the fascinating world of AI, how to craft effective prompts, and the importance of understanding AI's limitations.

As you continue your journey with AI, always remember to stay safe and protect your privacy. Treat AI as a helpful tool, not a friend, and be mindful of the information you share. And most importantly, never stop learning and exploring new ideas.

With your newfound AI knowledge, you're ready to delve deeper into the world of ChatGPT and create amazing conversations that help you learn, grow, and have fun. As you embark on your AI adventures, remember to stay curious and keep experimenting with different prompts and activities.

Happy chatting, and may your AI journey be filled with knowledge, creativity, and excitement!

100 prompts to get you going

1. What are the main differences between AI and human intelligence?
2. Can you explain the concept of machine learning in simple terms?
3. How do self-driving cars use AI to navigate safely?
4. How does AI like Siri or Alexa process voice commands?
5. Can AI create music or art? If so, how?
6. What are some famous AI robots and what can they do?
7. How is AI used in video games to create a more immersive experience?
8. What are some examples of AI used in healthcare?
9. Can AI predict the weather? How does it work?
10. How do AI language translators work?
11. What are some future applications of AI that we might see soon?
12. How can AI help protect the environment?
13. What is the Turing Test, and why is it important in AI research?
14. How do AI chatbots like ChatGPT learn from their mistakes?
15. Can AI be biased? If so, how can we prevent this?
16. What are some ethical concerns regarding AI development?
17. How do AI-powered recommendation systems work, like those on streaming services?
18. Can AI help solve complex math problems?
19. What role does AI play in space exploration?

20. How does AI assist in disaster response and recovery efforts?

21. How is AI used in the field of agriculture?

22. Can AI be used for language learning? How?

23. How does AI impact the job market?

24. How does AI help in detecting and preventing cyber attacks?

25. Can AI be used to solve global issues like poverty and hunger?

26. Describe the process of photosynthesis in plants.

27. Explain the water cycle and its importance.

28. What are the different states of matter, and how do they change?

29. How do magnets work, and what are their uses?

30. Explain the basics of electricity and how it powers our devices.

31. What is the difference between a solar and lunar eclipse?

32. How do volcanoes form, and why do they erupt?

33. Explain the process of plate tectonics and how it affects Earth's geography.

34. What are the different types of rocks, and how do they form?

35. How does the food chain work in an ecosystem?

36. Describe the life cycle of a frog.

37. What are the parts of a plant, and what are their functions?

38. How do sound waves travel, and how do we hear them?

39. What are the different types of clouds, and what do they indicate about the weather?

40. Explain the concept of gravity and its effects on Earth.

41. What are the key events and figures in American history?
42. Explain the importance of the Declaration of Independence.
43. What were the main causes of the American Civil War?
44. Who were some influential women in history, and what did they accomplish?
45. What are the seven wonders of the ancient world, and where are they located?
46. How did the invention of the printing press impact society?
47. Explain the significance of the Silk Road in world history.
48. What were the main achievements of ancient civilizations like Egypt, Greece, and Rome?
49. Who were some notable inventors and scientists throughout history, and what did they contribute?
50. Describe the events that led to World War I and World War II.
51. Can you recommend a good book for 5th graders?
52. Write a short story about a time-traveling adventure.
53. What are some interesting facts about famous authors?
54. Explain the elements of a good story.
55. What are some popular literary genres, and what makes them unique?
56. How can I improve my writing skills?
57. What are some tips for overcoming writer's block?
58. Can you provide a list of famous quotes from literature?
59. How do authors create compelling characters in their stories?
60. What are some creative writing prompts for practicing storytelling?

61. What are the rules for adding and subtracting fractions?

62. Explain the concept of decimals and their relationship to fractions.

63. How do you find the area of a rectangle, triangle, and circle?

64. What are prime numbers, and why are they important?

65. Explain the concept of ratios and how they are used in real-life situations.

66. What are the different types of angles, and how do you measure them?

67. How do you solve simple algebraic equations?

68. What is the Pythagorean theorem, and how do you use it?

69. Explain the concept of probability and how it is used in everyday life.

70. What is the order of operations, and why is it important in math?

71. Write a short poem about friendship.

72. What are some famous poems and poets throughout history?

73. Explain the different types of poetry and their characteristics.

74. How can I improve my poetry writing skills?

75. What are some common themes and symbols used in poetry?

76. How do poets use rhyme, rhythm, and other devices to create meaning in their work?

77. Write a haiku about nature.

78. What is the difference between a metaphor and a simile?

79. How do poets use imagery to create vivid pictures in the

reader's mind?

80. What is the purpose of a sonnet, and how is it structured?

81. What are the major organs in the human body, and what are their functions?

82. How does the circulatory system work?

83. Explain the process of digestion and the role of the digestive system.

84. What are the five senses, and how do they help us perceive the world around us?

85. How does the respiratory system function, and why is it important?

86. What are the different types of cells in the human body, and what are their functions?

87. Explain the function of the nervous system and how it communicates with the rest of the body.

88. How does the human body fight off infections and diseases?

89. What are the main components of a healthy diet, and why are they important?

90. Explain the importance of exercise and how it benefits the human body.

91. What are some effective strategies for studying and retaining information?

92. How can I improve my public speaking skills?

93. What are some tips for managing time and staying organized?

94. How can I develop good habits for success in school and life?

95. What are some strategies for setting and achieving goals?

96. How can I improve my listening and communication skills?

97. What are some tips for staying motivated and focused on my goals?

98. How can I build a strong support system and develop healthy relationships?

99. What are some ways to cope with stress and maintain a positive mindset?

100. How can I develop my leadership skills and make a positive impact on my community?

Kid-friendly AI tools and websites and how to use them

1. **Brainly** (https://brainly.com): Brainly is a social learning platform where students can ask questions, get homework help, and learn from their peers. Kids can ask questions, answer others' queries, and collaborate on various subjects.

How to use: Sign up for a free account, choose your subjects, and start asking or answering questions. Remember to follow the guidelines and be respectful of other users.

2. **Socratic by Google** (https://socratic.google.com): Socratic is an AI-powered mobile app designed to help students find answers to questions in various subjects, including math, science, and social studies.

How to use: Download the app on your smartphone or tablet, type in your question or take a photo of your homework problem, and the AI will help you find the best resources and explanations.

3. **QuillBot** (https://quillbot.com): QuillBot is an AI-powered paraphrasing tool that helps students improve their writing by suggesting better ways to phrase sentences and ideas.

How to use: Go to the website, enter your text, and click "Paraphrase." QuillBot will generate a rephrased version of your text while maintaining the original meaning.

4. **ALEKS** (https://www.aleks.com): ALEKS is an AI-powered adaptive learning platform that helps students learn math and science topics at their own pace. It identifies gaps in knowledge and provides personalized learning paths.

How to use: Sign up for an account, complete an initial assessment, and follow the customized learning plan to improve your skills.

5. **Wordtune** (https://www.wordtune.com): Wordtune is an AI-powered writing assistant that helps students improve their writing by offering suggestions on how to rephrase sentences and make them more concise, clear, and engaging.

How to use: Install the Wordtune extension for your web browser, and use it in your favorite writing tools like Google

Docs or Microsoft Word. Select a sentence, and Wordtune will offer suggestions for improvement.

6. **Blockly** (https://blockly.games): Blockly is a visual programming language designed for kids to learn coding concepts through puzzles and games. It uses blocks that represent programming concepts like loops, variables, and functions.

How to use: Visit the website, choose a game or puzzle, and start solving problems by connecting the blocks to create code.

7. **Duolingo** (https://www.duolingo.com): Duolingo is a fun and engaging language learning platform that uses AI to personalize the learning experience. It offers a variety of languages to choose from.

How to use: Sign up for a free account, select the language you want to learn, and start completing lessons, quizzes, and interactive exercises.

8. **Tynker** (https://www.tynker.com): Tynker is an online platform that teaches kids how to code through visual programming and game design. It offers a variety of coding courses and activities for different age groups.

How to use: Sign up for an account, choose a course or activity, and start learning to code with interactive lessons and projects.

9. **Khan Academy** (https://www.khanacademy.org): Khan Academy is a popular online learning platform that offers free video lessons, quizzes, and exercises on a wide range of subjects, including math, science, history, and more.

How to use: Create a free account, choose the subjects you're interested in, and start learning through videos and interactive exercises.

10. **Sumdog** (https://www.sumdog.com): Sumdog is a game-based learning platform that helps kids practice math, spelling, and grammar skills while having fun.

How to use: Sign up for a free account, choose the subject area you want to practice, and start playing games to improve your skills.

11. **Prodigy** (https://www.prodigygame.com): Prodigy is an engaging math game that uses an adaptive learning platform to help kids practice math skills while they embark on an adventure.

How to use: Sign up for a free account, create your character, and start completing math challenges in a fun, game-based environment.

12. **Quizlet** (https://quizlet.com): Quizlet is a study tool that helps students learn and memorize information through flashcards, quizzes, and games. It covers a variety of subjects

and is great for test preparation.

How to use: Sign up for an account, search for study sets created by other users or create your own, and start practicing with flashcards, quizzes, and games.

13. **Boddle** (https://boddlelearning.com): Boddle is a game-based math platform designed for elementary students to practice math skills while having fun in an immersive world.

How to use: Sign up for an account, select your grade level, and start playing math games that adapt to your learning pace and skill level.

14. **Code.org** (https://code.org): Code.org is a nonprofit organization that offers free coding courses, activities, and resources for kids to learn computer science concepts and skills.

How to use: Visit the website, choose a course or activity based on your age and interests, and start learning to code through engaging tutorials and projects.

15. **Scratch** (https://scratch.mit.edu): Scratch is a free, online programming platform developed by MIT that allows kids to create interactive stories, games, and animations while learning basic coding concepts.

How to use: Sign up for an account, explore the tutorials and resources, and start creating your projects by connecting blocks to create code.

These kid-friendly AI tools and websites provide engaging and interactive learning experiences, helping students improve their skills in various subjects. Remember to always follow safety guidelines and never share personal information while using these platforms.

16. **ChatGPT by OpenAI** (https://chat.openai.com): ChatGPT is an AI-powered chatbot developed by OpenAI. It can help students with various tasks like answering questions, providing explanations, brainstorming ideas, and offering writing assistance.

How to use: Visit the website and sign up for an account. Start typing your questions or prompts in the chatbox, and ChatGPT will generate a response. Remember to follow safety guidelines and never share personal information while using the platform.

List of future AI jobs you may find interesting

1. **AI Ethicist:** As an AI ethicist, you'll help companies and governments develop ethical guidelines and policies for the development and use of AI technologies, ensuring they are used responsibly and fairly.

2. **AI Trainer:** AI trainers work closely with AI systems to teach them new skills, correct errors, and improve their understanding. This role may involve creating training data, fine-tuning models, or teaching AI systems to understand context and nuances.

3. **AI Healthcare Specialist:** In this role, you'll work with AI-powered tools to analyze medical data, diagnose diseases, develop personalized treatment plans, and monitor patient progress, improving healthcare outcomes and efficiency.

4. **AI Product Designer:** As an AI product designer, you'll be responsible for designing user-friendly AI applications and experiences for various industries, ensuring that AI tools are accessible, useful, and engaging.

5. **AI Robotics Engineer:** AI robotics engineers design, develop, and program intelligent robots for various tasks, such as manufacturing, transportation, and healthcare. This role combines AI, mechanical engineering, and software development skills.

6. **AI Climate Analyst:** In this job, you'll use AI technologies to analyze climate data, predict climate changes, and develop solutions to mitigate the impact of climate change on our environment and society.

7. **AI Educator:** As an AI educator, you'll teach students and professionals about AI concepts, technologies, and applications. This role may involve developing AI curricula, teaching courses, or creating educational content.

8. **AI Content Creator:** In this creative role, you'll use AI-powered tools to generate unique content, such as articles, music, or art, pushing the boundaries of creativity and exploring new forms of expression.

9. **AI Urban Planner:** AI urban planners utilize AI technologies to analyze data, design smart cities, and improve urban infrastructure, ensuring that cities are more efficient, sustainable, and livable.

10. **AI Space Scientist:** As an AI space scientist, you'll work with AI technologies to analyze data from space missions, predict astronomical events, and develop solutions for space exploration and colonization.

11. **AI Agricultural Specialist:** In this role, you'll use AI tools to optimize farming practices, monitor crop health, and predict weather patterns, helping to improve crop yields and ensure food security.

12. **AI Legal Advisor:** AI legal advisors specialize in understanding the legal implications of AI technologies, helping companies navigate complex regulations and ensure that AI applications comply with laws and protect user privacy.

13. **AI Personalized Entertainment Designer:** In this creative role, you'll develop personalized entertainment experiences using AI, such as custom video games, interactive stories, or adaptive music playlists, tailored to individual preferences and emotions.

These future AI jobs offer exciting opportunities for children to consider as they grow up in a world increasingly driven by artificial intelligence. By exploring these careers, they can develop valuable skills and make a positive impact on society through the responsible and innovative use of AI technologies.

14. **AI Prompt Engineer:** As an AI prompt engineer, you'll specialize in crafting effective prompts to guide AI systems in generating useful and accurate responses. This role requires a deep understanding of how AI models work and strong communication skills to convey the desired information or task.

15. **AI Conversational Designer:** In this role, you'll design and develop natural and engaging interactions between AI chatbots or virtual assistants and users. This involves creating conversation flows, developing effective prompts,

and optimizing the AI's responses to ensure a seamless user experience.

Cool things AI would do in the future

1. **Personalized Education:** AI-powered learning platforms could offer personalized education for every student, adapting to their learning styles, strengths, and weaknesses, and providing a tailored curriculum that ensures effective learning and skill development.

2. **Smart Homes and Cities:** AI could play a major role in designing energy-efficient smart homes and cities that optimize resource usage, reduce waste, and improve the quality of life for residents.

3. **Advanced Healthcare:** AI could revolutionize healthcare by providing early and accurate diagnoses, personalized treatment plans, and drug discoveries, leading to better patient outcomes and potentially saving millions of lives.

4. **Autonomous Vehicles:** AI-powered self-driving cars could become the norm, reducing traffic congestion, improving road safety, and providing efficient transportation options for everyone, including the elderly and those with disabilities.

5. **AI-Assisted Creative Arts:** AI could help artists, musicians, and writers explore new forms of creative expression by generating unique ideas, assisting in the creative process,

and even creating original works of art, music, or literature.

6. **Environmental Conservation:** AI could aid in predicting natural disasters, monitoring and protecting endangered species, and optimizing resource consumption, helping to preserve our planet and mitigate the effects of climate change.

7. **Space Exploration:** AI-powered robots and data analysis tools could greatly enhance space exploration, aiding in the discovery of new celestial bodies, the search for extraterrestrial life, and the development of advanced space travel technologies.

8. **AI Personal Assistants:** Advanced AI personal assistants could become indispensable partners in daily life, helping with scheduling, communication, decision-making, and even emotional support, making our lives more organized and stress-free.

9. **Advanced Robotics:** AI could lead to the development of sophisticated robots capable of performing complex tasks, such as medical surgeries, search and rescue operations, and disaster relief, ultimately saving lives and improving safety.

10. **AI-Enabled Virtual Reality:** AI technologies could enable immersive and adaptive virtual reality experiences, allowing users to explore simulated worlds, learn new skills, and connect with others in ways never before possible.

These cool things AI could do in the future showcase the potential for artificial intelligence to revolutionize various aspects of our lives, from education and healthcare to creative arts and environmental conservation. As AI technologies continue to advance, they hold the promise of making the world a better, safer, and more efficient place for all.

Cool things AI can help you with as a 5th grader

1. **Homework Assistance:** AI-powered tools like Brainly or Socratic can help you find answers and explanations to your homework questions, making it easier to understand and complete assignments.

2. **Writing Support:** AI writing assistants like QuillBot and Wordtune can help you improve your writing by suggesting better ways to phrase your sentences and making your ideas clearer and more engaging.

3. **Math Practice:** AI-driven math platforms like Prodigy and Sumdog provide engaging games and exercises to practice your math skills, making learning more fun and interactive.

4. **Learning Languages:** AI-powered language learning apps like Duolingo offer personalized lessons and practice exercises, making it easier to learn and practice a new language.

5. **Reading Comprehension:** AI tools can help you improve your reading skills by offering reading recommendations, highlighting important information, and even providing summaries of complex texts.

6. **Study and Memorization:** AI-powered study tools

like Quizlet can help you learn and memorize information more effectively through flashcards, quizzes, and games, making it easier to prepare for tests.

7. **Coding Skills:** AI platforms like Blockly, Tynker, and Scratch teach coding concepts through puzzles and games, making it fun and engaging to learn programming skills.

8. **Creative Expression:** AI-powered creative tools can help you explore your artistic side, assisting with art projects, music composition, and storytelling, inspiring new ideas and innovative creations.

9. **Time Management:** AI personal assistant apps can help you manage your time more effectively, keeping track of assignments, setting reminders, and helping you prioritize tasks.

10. **Social Skills and Emotional Support:** AI chatbots can provide a safe space for practicing communication, offering advice on social situations, and even providing emotional support when you need someone to talk to.

These cool things AI can help with show the potential for artificial intelligence to support your learning and personal growth as a 5th grader. By leveraging these tools, you can enhance your education, develop new skills, and enjoy a more engaging and interactive learning experience.

Fun games and interactive activities

(For the reader, parent, or educator)

AI Prompt Scramble:

1. Create a word bank with different subjects, themes, and situations. Have students randomly pick a combination of words and use them to craft unique prompts for ChatGPT. They can share their prompts with classmates and compare the AI-generated responses.

AI Story Chain:

2. Begin a story with an opening line, and have students take turns adding sentences using ChatGPT's suggestions. This collaborative activity encourages creativity and teamwork while demonstrating the AI's ability to generate context-aware content.

Emoji Code Breaker:

3. Develop a list of sentences or phrases represented by emojis. Students must decipher the emoji code and input the corresponding text as a prompt to ChatGPT, which should generate a response based on the decoded message. This activity encourages problem-solving skills and creative thinking.

AI Art Collaboration:

4. Students can provide a prompt describing a scene or an object to ChatGPT. Based on the AI-generated response, they

will create a drawing or painting inspired by the description. This activity fosters artistic expression and showcases the AI's ability to inspire creativity.

ChatGPT Trivia:

5. Create a list of trivia questions for students to ask ChatGPT. They can work individually or in teams to find the answers using the AI, encouraging research skills and critical thinking.

AI Songwriter Challenge:

6. Have students come up with a theme or topic for a song. They can then use ChatGPT to generate lyrics or song ideas, working together to compose an original tune. This activity highlights the AI's potential for creative applications.

AI-powered Debate:

7. Divide students into teams, assign a topic, and have them use ChatGPT to generate arguments for or against the topic. They can then present their AI-assisted arguments in a debate, promoting public speaking and critical thinking skills.

ChatGPT's Word Ladder:

8. Students can provide a starting word and an ending word, then use ChatGPT to create a word ladder – a series of words that change by only one letter at a time. This activity encourages vocabulary development and problem-solving.

AI Career Exploration:

9. Students can use ChatGPT to explore various AI-

related careers by asking questions about job responsibilities, required skills, and potential future developments. This activity exposes them to the exciting opportunities that AI offers in the job market.

AI Time Capsule:

10. Students can create a list of predictions about the future of AI and technology. Using ChatGPT, they can explore the plausibility of their predictions and generate AI's perspective on future developments. This activity fosters critical thinking and curiosity about the potential impact of AI on society.

These interactive activities and games are designed to engage students, encourage creativity, and demonstrate the diverse applications of AI, making the learning process fun and enjoyable.

Definitions

1. Artificial Intelligence (AI): A field of computer science that focuses on creating machines and software that can think, learn, and make decisions like humans do.

2. Chatbot: A computer program that can have a conversation with people, either through text or spoken words, by answering questions or providing information.

3. Algorithm: A set of step-by-step instructions that a computer follows to solve a problem or complete a task.

4. Machine Learning: A type of artificial intelligence where computers learn from data and experiences without being explicitly programmed, allowing them to improve their performance over time.

5. Natural Language Processing (NLP): A field of artificial intelligence that focuses on helping computers understand, interpret, and respond to human language.

6. Prompt: A question, statement, or instruction that guides an AI's response or action.

7. Circuit: A closed path through which electricity flows, typically found in electronic devices and machines.

8. Autonomous: Able to operate or make decisions

independently, without direct human control or supervision.

9. Robotics: A field of engineering that deals with designing, building, and programming robots – machines that can perform tasks or actions on their own.

10. Virtual Reality (VR): A technology that creates a computer-generated environment that people can explore and interact with, often using special headsets or gloves.

11. Algorithm Bias: A situation where an AI system's outputs, like decisions or predictions, unfairly favor certain groups or individuals based on factors like race, gender, or age, often due to biased data used for training the AI.

12. Data Science: A field that combines computer science, statistics, and domain knowledge to analyze and interpret large amounts of data, helping to make informed decisions and predictions.

13. Ethicist: A person who studies ethics, which involves understanding what is right and wrong and how to make good choices.

14. Diagnostic: A process or test used to determine the cause of a problem, illness, or condition.

15. Extraterrestrial: Something that comes from or exists outside Earth, often used to describe the possibility of life on other planets.

www.ingramcontent.com/pod-product-compliance
Lightning Source LLC
Chambersburg PA
CBHW042347030426
42335CB00031B/3492